Using Simple Machines

Inclined Planes All Around

by Trudy Becker

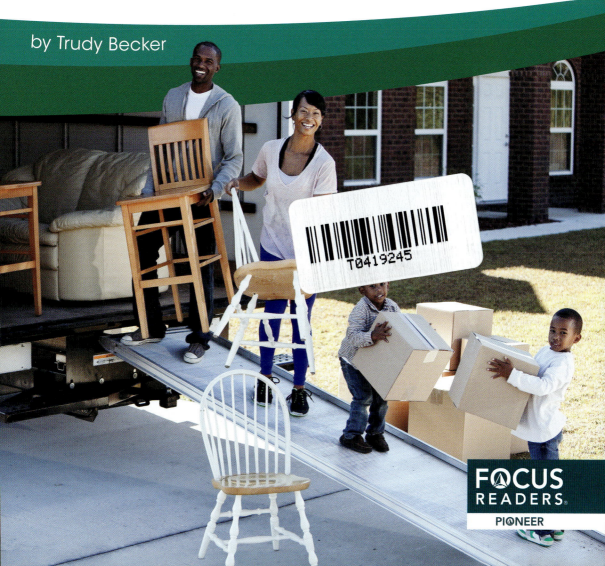

FOCUS READERS®
PIONEER

www.focusreaders.com

Copyright © 2024 by Focus Readers®, Lake Elmo, MN 55042. All rights reserved. No part of this book may be reproduced or utilized in any form or by any means without written permission from the publisher.

Focus Readers is distributed by North Star Editions:
sales@northstareditions.com | 888-417-0195

Produced for Focus Readers by Red Line Editorial.

Photographs ©: iStockphoto, cover, 1, 8; Shutterstock Images, 4, 6, 10, 12, 14 (top), 14 (bottom), 17, 18, 20

Library of Congress Cataloging-in-Publication Data
Names: Becker, Trudy, author.
Title: Inclined planes all around / by Trudy Becker.
Description: Lake Elmo, MN: Focus Readers, [2024] | Series: Using simple machines | Cover title. | Includes bibliographical references and index. | Audience: Grades K-1
Identifiers: LCCN 2022058386 (print) | LCCN 2022058387 (ebook) | ISBN 9781637395974 (hardcover) | ISBN 9781637396544 (paperback) | ISBN 9781637397664 (ebook pdf) | ISBN 9781637397114 (hosted ebook)
Subjects: LCSH: Inclined planes--Juvenile literature.
Classification: LCC TJ1428 .B43 2024 (print) | LCC TJ1428 (ebook) | DDC 621.8--dc23/eng/20230105
LC record available at https://lccn.loc.gov/2022058386
LC ebook record available at https://lccn.loc.gov/2022058387

Printed in the United States of America
Mankato, MN
082023

About the Author

Trudy Becker lives in Minneapolis, Minnesota. She likes exploring new places and loves anything involving books.

Table of Contents

CHAPTER 1
Moving Day 5

CHAPTER 2
What Are Inclined Planes? 9

CHAPTER 3
Ramps Everywhere 13

THAT'S AMAZING!
Dump Trucks 16

CHAPTER 4
Fun with Inclined Planes 19

Focus on Inclined Planes • 22
Glossary • 23
To Learn More • 24
Index • 24

Chapter 1

Moving Day

A boy carries a heavy box. He takes it to a moving truck. It's hard to lift the box up high. So, he walks up the **loading ramp** instead. He gets the box into the truck.

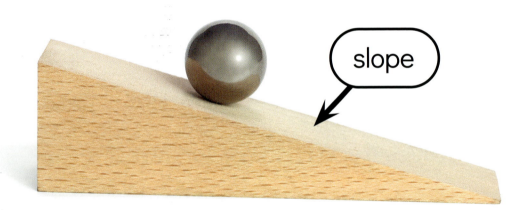

The loading ramp is an inclined plane. That means one end is higher than the other. It has a **slope**. Inclined planes are one of the six **simple machines**.

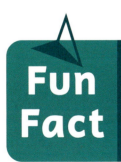

Fun Fact Big hills can be inclined planes. Sometimes people roll down the slopes.

Chapter 2

What Are Inclined Planes?

All simple machines help people do jobs. People use inclined planes for many reasons. They can help people move things. The inclined planes make it easier.

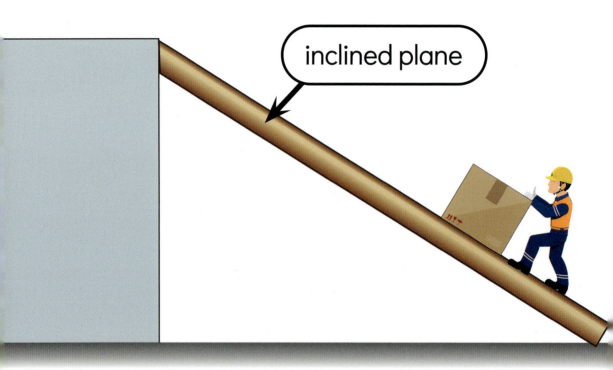

Lifting items straight up is hard. But pushing or pulling with a ramp can help. Less **force** is needed. The items move **vertically** and **horizontally** at the same time.

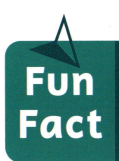

Fun Fact Inclined planes can help move objects down, too.

Chapter 3

Ramps Everywhere

Inclined planes are all around. Many buildings have wheelchair ramps. People who use wheelchairs can use the ramps. That helps them get in and out. They can avoid stairs.

gangway

Many roads and paths are inclined planes, too. Mountain roads help people reach the top. Boats can have sloped **gangways**. That helps people get on.

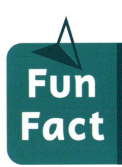

Fun Fact Sometimes the roof of a house is an inclined plane. Rain can slide off.

15

That's Amazing!

Dump Trucks

Dump trucks use inclined planes. First, the truck bed is filled. It can hold a lot of trash. Then, the bed rises up. It makes a slope. The trash slides down. People don't have to carry the trash out.

Chapter 4

Fun with Inclined Planes

Inclined planes aren't just for jobs. People can use them for fun. Slides are inclined planes. If the slope is **steep**, people can rush down quickly.

Skate parks have inclined planes, too. Skaters zoom up and down ramps. That helps the skaters go high. They gain speed on the way down.

Fun Fact Roller coasters have inclined planes. Steep slopes make the rides exciting.

FOCUS ON
Inclined Planes

Write your answers on a separate piece of paper.

1. Write a sentence that explains the main idea of Chapter 2.

2. What is the most helpful way you use inclined planes in your life? Why?

3. What is an inclined plane outside of many buildings?
    - A. outdoor elevator
    - B. hand railing
    - C. wheelchair ramp

4. When objects move along an inclined plane, what direction do they go?
    - A. vertically and horizontally
    - B. only vertically
    - C. only horizontally

Answer key on page 24.

Glossary

force
A push or pull that changes how something moves.

gangways
Narrow walkways to get onto or off of something.

horizontally
Side to side.

loading ramp
A ramp used to move things into or out of something.

simple machines
Machines with only a few parts that make work easier.

slope
A surface with one end higher than the other.

steep
Much higher on one end than the other.

vertically
Up and down.

To Learn More

BOOKS

Blevins, Wiley. *Let's Find Inclined Planes*. North Mankato, MN: Capstone Press, 2021.

Mattern, Joanne. *Inclined Planes*. Minneapolis: Bellwether Media, 2020.

NOTE TO EDUCATORS

Visit **www.focusreaders.com** to find lesson plans, activities, links, and other resources related to this title.

Index

F
force, 11

R
ramps, 5, 7, 11, 13, 21

S
slopes, 7, 15, 16, 19, 21

T
trucks, 5, 16

Answer Key: 1. Answers will vary; 2. Answers will vary; 3. C; 4. A